# Algebra Book

This book is protected by copyright and
cannot be reproduced without permission

Jetser Carasco

# Table of contents

1. Integers  ------------------------- 3

2. Adding/subtracting integers ---- 5

3. Multiplying/dividing integers --15

4. Solving equations --------------- 23

5. Solving/graphing inequalities - 47

6. Exponent/roots/scientific notation/ the Pythagorean Theorem ------ 53

# Unit 1: Integers

Main tool: a number line

A number line is a great tool that can help anyone master integers quickly.

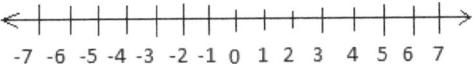

Take a look at the number line above while some important terms are explained.

*Natural numbers*: 1, 2, 3, 4, 5, and so forth

*Whole numbers*: 0, 1, 2, 3, 4, 5, and so forth.

*Opposites*: numbers located at the same distance from zero.

For example, 4 is 4 units away from 0 and -4 is 4 units away from 0, so 4 and –4 are opposites.
Similarly, 5 and -5 are opposites.
Notice that zero has no opposite.

*Integers*: whole numbers and their opposites. The number line above has all integers from -7 to 7

Positive integers are numbers to the right of 0 and are read positive 1 or + 1, positive 2 or +2, and so on.
Zero is neither positive nor negative.

Negative integers are numbers to the left of 0 and are read negative 1 or -1, negative 2 or -2, and so on.

Zero is neither positive nor negative.

3

## Absolute value

The absolute value is the distance a number is from 0. Looking at the graph below, we see that the number 2 is two units away from 0, so absolute value of 2 is 2. We write | 2 | = 2

-3 is three units away from 0, so absolute value of -3 is 3. We write | -3 | = 3

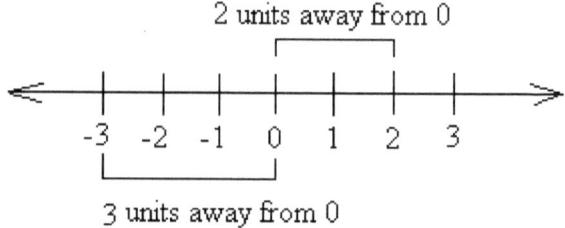

## Comparing integers

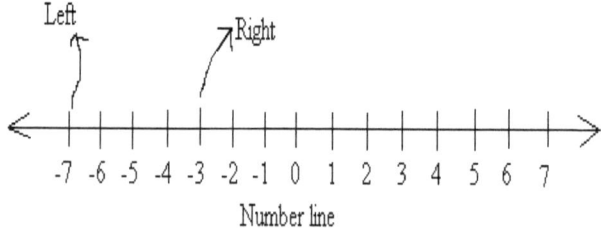

Number line

**Key concept**: a number on the right is always bigger than a number on the left.

For example, -3 is on the right of -7, so -3 is bigger than -7. We write -3 > -7

You could also say that -7 is smaller than -3 and write -7 < -3

# Adding/Subtracting integers

Adding integers with a number line is straightforward!

**Operations**          **Signs**
Addition: +             Positive (+)
Subtraction (—)         Negative (-)_

**Key concept #1**: place the number on the left on the number line and either move to the right or to the left depending on the operation or the sign of the number on the right.

**Key concept #2:** when a number is positive, you can omit the + sign!
Thus +5 = 5

**Cases**

Case #1: adding a positive

When adding a positive number to the number on the left as in -4 + 5 = -4 + +5
or 4 + 2 = 4 + +2, move to the right

Case #2: adding a negative

When adding a negative number to the number on the left as in -2 + -5 or 2 + -5, move to the left.

Case #3: subtracting a positive

When subtracting a positive number from the number on the left as in -4 — 2 = -4 — +2 or 3 — 2 = 3 — +2, move to the left.

Case #4: subtracting a negative

When subtracting a negative number to the number on the left as in -4 — -2 or
3 — -2 = 3 — -2, move to the right.

Why do we move to the right? Great question! You can either move to the right or to the left. When you subtracted a positive number as in case # 3, you moved to the left, so if you are subtracting a negative number you cannot move to the left again. You have to move to the right.

**Example # 1:** taken from case #1

Add -4 and 5. This means the same as -4 + 5 or -4 + + 5.

The number on the left is -4, so put -4 on the number line. Then, move 5 units to the right

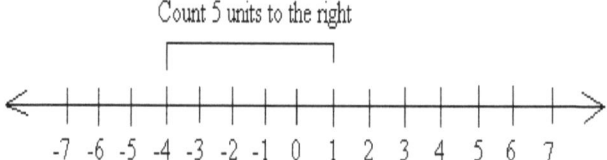

Since you stop at 1, -4 + 5 = 1

**Example # 2:** taken from case #2

Add 2 and -5. This means the same as 2 + -5.

The number on the left is 2, so put 2 on the number line. Then, move 5 units to the left.

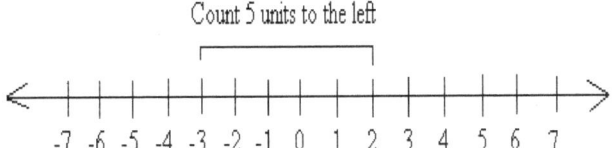

Since you stop at -3, 2 + -5 = -3

**Example # 3:** 2 — 5 or 2 — +5

The number on the left is 2, so put 2 on the number line. Then, move 5 units to the left as explained in case #3. Since you are doing the exact same thing as example #2, you will get the same answer as example #2.

2 + -5 = 2 — 5 = 2 — +5 = -3

We then have the following concept.

**Key concept #3:** adding the negative give the same answer as subtracting the positive.(— = + -)

**Example # 4:** -4 — -5

The number on the left is -4, so put -4 on the number line. Then, move 5 units to the right as explained in case #4. Since you are doing the exact same thing as example #1, you will get the same answer as example #1.

7

$-4 - -5 = -4 + 5 = -4 + +5 = 1$

We then have the following concept

**Key concept #4:** subtracting the negative give the same answer as adding the positive.($- - = + +$)

**Key concept #5:** adding opposite numbers always equals to 0

$-6 + 6 = 0$
$-100 + 100 = 0$
$-244 + 244 = 0$

Concept #5 is especially useful when you are adding big integers.

**Example # 5:** $-244 + 246$.

Move to the right (concept #1). Add 244 to -244 to get to 0 (concept #5). You have an extra of 2 units ($246 - 244 = 2$), so you will be on the positive side of the number line at 2. You could also do it another way.

$-244 + 246 = -244 + 244 + 2 = 0 + 2 = 2$

$-244 + 246 = 2$

**Example # 6:** $-244 + 243$.

Move to the right (concept 1). You need 244 to get to 0 (concept 5). You only have 243. You are short by 1. So you will be on the negative side of the number line at -1.

$-244 + 243 = -1$

## Observation #1:

**Example 5 gave a positive answer while example 6 gave a negative answer**

When the numbers have different sign, how do you know when the answer is positive, or negative?

Concept #5 is crucial here because it tells you what you need to add to get to 0. In example # 5, you only needed 244 to get to 0, yet you have 246, so you will pass 0 and be on the positive side.

On the other hand, in example 6, you did not have 244 to get to 0; you had 243. Since you never reach zero, you are on the negative side.

**Example # 7:** -244 — -243.

Since -244 — -243 = -244 + 243, this is the same problem as example # 6.

-244 — -243 = -1

**Example # 8:** -3 + -2

The number on the left is -3, so put -3 on the number line. Then, move 2 units to the left as explained in case #2.

You stopped at -5, so -3 + -2 = -5

**Example # 9:** 2 + 4

The number on the left is 2, so put 2 on the number line. Then, move 4 units to the right as explained in case #1.

You stopped at 6, so 2 + 4 = 6

## Observation #2:

For example # 8 and example # 9, the answer has the same sign as the numbers.
Thus, when the numbers have the same signs, pretend the signs are not there when you add. After you get the sum, put the sign next to the answer.

**Example # 10:** -200 + -100

The numbers have the same sign, so add 200 and 100 and put a negative sign next to the answer. -200 + -100 = -300.

**Example # 11:** 200 − -100

200 − -100 = 200 + 100 = + 200 + +100.

Again, the numbers have the same sign, so add 200 and 100 and put a positive sign next to the answer. 200 +100 = +300.

## Summary and key exercises

The best way to tackle any integer problems when adding or subtracting is to use a number line. Put the number on the left on the number line and decide if you should move to the right or to the left.

**Exercises**
For 1), 2), 3), 4), and 5, you move to the right. For 6), 7), 8), and 9), you move to the left.

1)    5 + 2 = 7

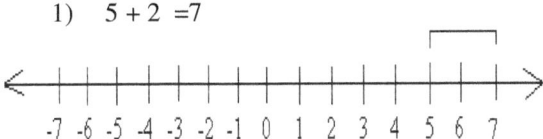

2)    5 – -2 = 7

3)    -5 + 2 = -3

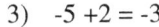
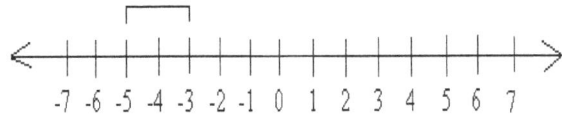

4)    -5 – -2 = -3

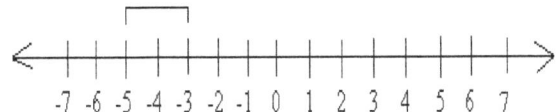

5)  -5 + 9 = 4

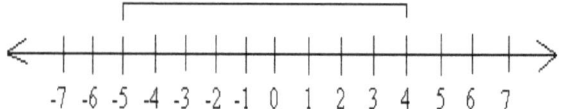

6)  6 − 2

7)  6 +-2

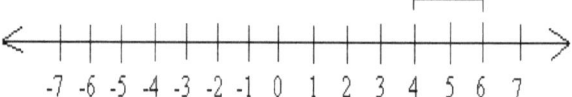

8)  -4 − 3

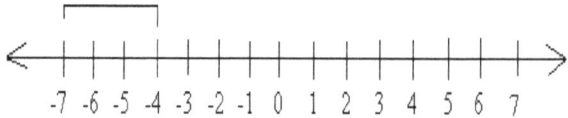

9)  -4 + -3

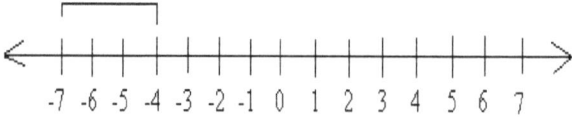

10) -221 + 200

Put -221 on the number line and move to the right. To get to 0, you need to add 221 to -221. Since you are only adding 200 and are not reaching 0, the answer is negative.

Where are you on the negative side? Well how much is missing to get to 221? 200 + what equal 221? It is 21. Therefore, you are at -21.

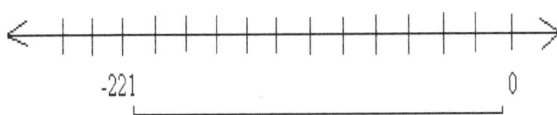

11) -221 + -200

Put -221 on the number line and go to the left. Since I was already on the negative and you are still going to the left, I am still on the negative side.

Thus, the answer is negative. Where are you on the negative side? 200 + 221 = 421. So you are at -421.

12) 350 + -200

Put 350 on the number line and go to the left. To get to 0, you need to add -350 or move 350 units to the left. Since you will move only 200 units to the left you are still on the positive side of the number line. Where are you?

Well, you still need to move 150 units to the left to get to 0, so you are at 150.
Thus, the answer is +150

13) 350 + -360

Put 350 on the number line and go to the left.
To get to 0, you need to add -350 or move 350 units to the left.

After you have moved 350 units to the left, any more moves will take you to the negative side of the number line.

How many more units can you still move? You can still move 10 units (360 – 350)
In the end, you will be at -10.
Thus, 350 + -360 = -10

## Multiplying/Dividing integers

Repeated addition is the key topic to understand multiplication of integers.

Multiplication is repeated addition and it is a fast way of adding the same number many times. For instance, instead of adding 2 together 3 times, you can just multiply 2 by 3 or 3 by 2.

$2 + 2 + 2 = 2 \times 3$

Since $2 + 2 + 2 = 6$ and 6 is positive, then it is easy to draw the following conclusion:

$2 \times 3 = +2 \ \times \ +3 = +6$

So, $+ \ \times \ + = +$

By the same fashion,

$2 + 2 + 2 = 3 \times 2$

$3 \times 2 = +3 \ \times \ +2 = +6$

And again, $+ \ \times \ + = +$

Now, look at the following repeated addition:  $-2 + -2 + -2$

Instead of adding -2 together 3 times, you can just multiply -2 by 3 or 3 by -2

$-2 + -2 + -2 = -2 \times 3$

Or

$-2 + -2 + -2 = 3 \times -2$

Now, $-2 + -2 + -2 = -4 + -2 = -6$

Thus, $-2 \times 3 = -2 \times +3 = -6$

So, $- \times + = -$

And $3 \times -2 = +3 \times -2 = -6$

So, $+ \times - = -$

So far, here is what we have:

$+ \times + = +$
$- \times + = -$
$+ \times - = -$

The last case is $- \times - = ?$

This one is a little subtle and cannot be seen with repeated addition.

However, we can use a simple pattern to see what a negative times a negative is.

$3 \times -5 = -15$

$2 \times -5 = -10$

$1 \times -5 = -5$

$0 \times -5 = 0$

$-1 \times -5 = ?$

**Carefully** examine the pattern above.

Can you predict what $-1 \times -5$ equal to?

If you could not make a good prediction, then ask yourself, "How do we get from -15 to -10, from -10 to -5, and from -5 to 0 ?"

Just add 5.

Thus, adding 5 to 0 equals 5.

$-1 \times -5 = 5$   Or

$-1 \times -5 = +5$

So $- \times - = +$


Putting it all together, we have:

$+ \times + = +$
$- \times\ + = -$
$+ \times\ - = -$
$- \times - = +$

Repeated addition will not help us to understand division of integers, but we can use facts that we already established for multiplication of integers to make logical conclusion for division of integers.

## Key concept #6:

If b is not equal to 0 and $\frac{a}{b} = c$, then

$a = b \times c$

You can also write $\frac{a}{b} = c$ as $a \div b = c$

For example, look at the following division problem:

$60 \div 5 = 12$

$60 \div 5 = 12$ because $5 \times 12 = 60$

By the same fashion,

$10 \div 2 = 5$

$10 \div 2 = 5$ because $2 \times 5 = 10$

We know 10 and 2 are positive, so the answer 5 has to be positive also.

$+10 \div +2 = +5$

Pretend for a second that 5 is negative

$+10 \div +2 = -5$

This would mean that $+2 \times -5 = +10$

And that contradicts something we already Established as true that is $+ \times - = -$

Therefore, $+10 \div +2 = +5$

We draw the following conclusion:

$+ \div + = +$

We can use the same logic to find out what a positive divided by a negative is and vice versa.

$+10 \div -2 = -5$

Notice that indeed $-2 \times -5 = +10$

We draw the following conclusion:

$+ \div - = -$

Will it contradict already established fact to say that $+10 \div -2 = +5$?

Yes, it will since $-2 \times +5 \neq +10$

The symbol $\neq$ means not equal to

In a similar way, $-10 \div +2 = -5$

Indeed, $+2 \times -5 = -10$

Again, saying that $-10 \div +2 = +5$ would not make sense.

$+2 \times +5 = 10$ while the answer should be $-10$

Since -10 ÷ +2 = -5,

We draw the following conclusion:

- ÷ + = -

Finally, -10 ÷ -2 = +5

The logic is still the same. It the answer was -5, then -2 × -5 will give 10, not -10.
Thus, 5 must be positive.

So, - ÷ - = +

We now have last key concept for this unit

## Key concept #7

| Multiplication | Division |
|---|---|
| + × + = + | + ÷ + = + |
| - × + = - | + ÷ - = - |
| + × - = - | - ÷ + = - |
| - × - = + | - ÷ - = + |

## Summary and key exercises

Notice that multiplication and division give the same answer.

Notice also that when the signs are the same, the answer is +. When the signs are different, the answer is -

**Exercises**: For 1) though 5) we see the numbers have different signs, so the answer is negative. Exercises 6) through 10 have the same sign, so the answer is positive.

1) -5 × 10 = -50

2) 10 × -10 = -100

3) -2 × + 4 = -8

4) +5 × -4 = -20

5) 6 × -2 = -12

6) -5 × -10 = 50

7) +5 × +10 = 50

8) 5 × 10 = 50

9) -8 × -10 = 80

10) -7 × -7 = 49

### _____Integers quiz_____

1) -10 + 9    _____

2) -12 + 14   _____

3) -15 ÷ 5    _____

4) -12 * 2    _____

5) - 10 – 9   _____

6) -12 × - 2  _____

7) -14 – -4   _____

8) -14 – -19  _____

9) 8 ÷ - 2    _____

10) -16 ÷ -4  _____

## Answers

### _____Unit 1 quiz_____

1) -1   2) 2   3) -3   4) -24   5) -19

6) 24   7) -10   8) 5   9) -4   10) 4

# Unit 2: Solving equations

Main tool: Algebra tiles

Here is the model and principles:

A *red square* always represents a negative number.

So, one red square is -1 and three red squares is -3.

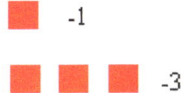

A *blue square* always represents a positive number.

So one blue square is +1 and 4 blue squares is + 4

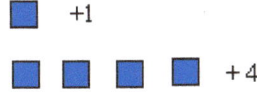

A blue square next to a red square always equals to 0.

An *equation* is anything with an equal sign.

The following are equations:

1) x = 4 + 1

2) -3x + 5 = 7

3) x + 1 = -2

4) a + 0 = y

5) 4x + 5y = 8

The following are **not** equations. They are called *expressions*.

1) 2 + 6

2) x + 8

3) 2z

4) 9 × 4 + 2

1) and 4) are called *numerical expressions*.

2) and 3) are called *algebraic expressions*.

What is the difference? 2) and 3) have variables x and z.

We called expressions with variables algebraic expressions

A *variable* is a letter that is used to represent an unknown number

Rectangles are used to represent variables

A blue rectangle represents + x

A red rectangle represents –x

An equal sign separates the left side and the right side of an equation.

$$\text{Left side} = \text{Right side}$$

For example, what do you see on the left side and the right side of the following equation?

$2x + 2 - 4 = 0$

On the left side, we see $2x + 2 - 4$.
On the right side, we see 0.

This concept is <u>extremely important</u> because we will be talking about right side and left all the time when solving equations.

It is also important to keep an equation balanced. To keep an equation, do the exact same thing to both sides.
For example, if you add 2 to the left side, add also 2 to the right side.

If you subtract 2 to a side, subtract 2 also to the other side.

We are ready to solve equations.

Solving an equation is the process that **isolates** the blue rectangle from either the left side or the right side.

1) x + 3 = 2

Model the equation with tiles.

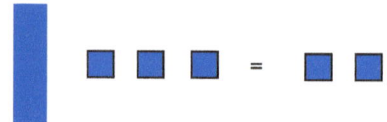

Put 3 red squares on the left side to isolate the blue rectangle. And to keep the equation balanced, put also 3 red squares on the right side as well.

**Common pitfall:** Putting 3 red squares on the left and not putting 3 red squares on the right.

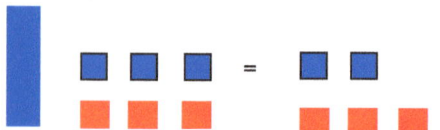

Remember that a red square always cancels with a blue square. Thus, cancel all zeros!

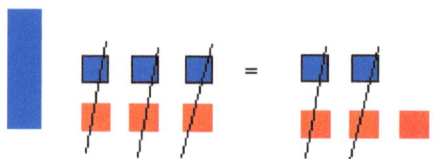

26

After all cancellations, you end up with the model below.

Translating the model above into an equation gives

x = -1

Let's take a look again at the following model

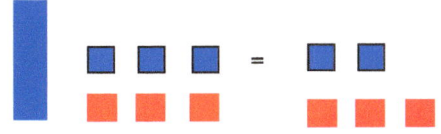

In general, it is simpler to write

x + 3 = 2
   -3     -3

Of course, 3 and -3 cancel and 2 and -3 equals -1 resulting in x = -1

2) x − 2 = -5

Here we need key concept #3 (− = + -)

So rewrite x − 2 = -3 as x + -2 = -3 and model the latter.

x + -2 = -5

Model the equation with tiles.

Put 2 blue squares on the left side to isolate the blue rectangle. And to keep the equation balanced, put also 2 blue squares on the right side as well.

Cancel all zeros.

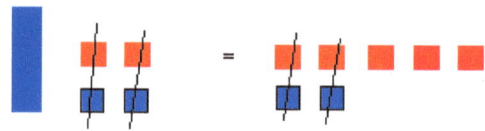

After all cancellations, you end up with the model below.

Translating the model above into an equation gives

$x = -3$

Again, in general, it is simpler to write

$$x + \frac{-2}{2} = \frac{-5}{2}$$

-2 and 2 cancel and -5 and 2 equals -3 resulting in

$x = -3$

We will make a slight change to 2) by putting a negative next to x.

3) $-x - 2 = -5$

The only difference is that instead of putting a blue rectangle put a red rectangle.

After you do the **exact same thing** for 2), you will end up with the following:

Since you are supposed to isolate a blue rectangle, all you need to do is to change the red rectangle into a blue rectangle.

And again, whatever you do to a side, you have to do it to the other side. Therefore, you will also change all 3 red squares on the right to 3 blue squares.

We get:

Translating the model above into an equation gives

x = 3.

It is simpler to write

-x − 2 = -5   or   -x + -2 = -5

-x + -2 = -5
  + 2     +2
_____

-x    = -3

How do you change -x into an x?
Multiply -x by a negative. Remember to do the same thing to the other side.

- × -x     = -3 × -

   x    =   3


4) 3x = -15

Model the equation with tiles.

The trick here is to see that since you are solving for x, all you need is to pull out one big blue rectangle.

Also, notice that each rectangle can be paired up with only 5 red squares. Another way to understand this is to pretend that the 3 blue rectangles are human beings and they will eat the red squares. To do this evenly, each rectangle should have only 5 red squares to eat.

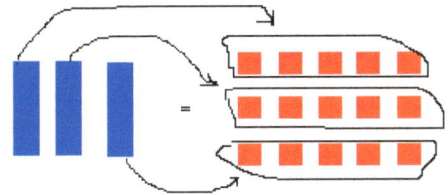

We get in this case:

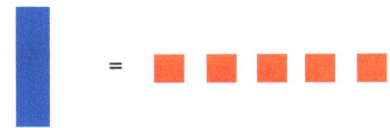

Translating the model above into an equation gives

x = -5

In example 4, you were basically sharing 15 red squares between 3 people. The way to do this is to divide 15 red squares by 3.

Thus, it is simpler to divide both sides by 3 and write.

$$\frac{3x}{3} = -\frac{15}{3}$$

1x = -5

x = -5

5) $\frac{x}{4} = 2$

This time, x is divided by 4, so break x down into 4 equal pieces and 1 piece is x/4 as shown below.

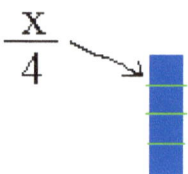

So, your model for this equation is the following:

■   =   ▫ ▫

Now, ask yourself, "How do I turn $\frac{1}{4}$ x to x?"

Copy the model 3 times and you will have $\frac{1}{4}$ x four times and when you have $\frac{1}{4}$ x four times, it is x!

Just connect the 4 pieces until there is no space between them.

The resulting shape is x.

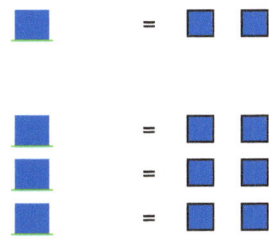

The 4 pieces on the left equals to x and the right side equals to 8

So x = 8

There is an easier way. Knowing that $\frac{1}{4}$ x times 4 equals x, we can just multiply both sides of the equation by 4.

$$\frac{1}{4}x = 2$$

$$4 \times \frac{1}{4}x = 4 \times 2$$

$$\frac{4}{1} \times \frac{1}{4}x = 4 \times 2$$

$$\frac{4}{4}x = 8$$

1x = 8

x = 8

6) $\dfrac{3x}{4} + 3 = 9$

Instead of modeling with tiles, let us use some logic and exercises 1), 4), and 5)

    1)   $x + 3 = 2$

In exercise 1) shown above we subtracted 3 from both sides of $x + 3 = 2$.

We can start by doing the same thing for

$$\dfrac{3x}{4} + 3 = 9$$

We can subtract 3 from both sides

$$\dfrac{3x}{4} + 3 - 3 = 9 - 3$$

$$\dfrac{3x}{4} = 6$$

Now, what you have here is a combination of exercise 4) and exercises 5).

Before we proceed, let us make a useful substitution.

Let $y = 3x$.

$$\dfrac{y}{4} = 6$$

In exercise 4) you had 3x =15 and

In exercise 5) you had $\dfrac{x}{4} = 2$

To solve exercise 5) you multiply both sides by 4. You will do the same thing here.

Multiply both sides of $\dfrac{y}{4} = 6$ by 4

$$4 \times \dfrac{y}{4} = 6 \times 4$$

$$y = 24$$

Now, replace y with 3x as it was before.

$$3x = 24$$

To solve exercise 4) you divided both sides by 3. You will do the same thing here.

Divide both sides of $3x = 24$ by 3.

$$\dfrac{3x}{3} = \dfrac{24}{3}$$

$$x = 8$$

You could do all these faster.

Multiplying by 4 and dividing what you get by 3 is the same as multiplying by $\dfrac{4}{3}$

Why? Let z be any number.

Multiplying z by 4 gives 4z. Then, dividing 4z by 3 gives $\dfrac{4z}{3}$

Now, just multiply z by $\dfrac{4}{3}$ to see if we get $\dfrac{4z}{3}$

$$z \times \dfrac{4}{3} = \dfrac{z}{1} \times \dfrac{4}{3} = \dfrac{z \times 4}{1 \times 3} = \dfrac{4z}{3}$$

So to solve $\dfrac{3x}{4} = 6$, you can just multiply both sides by $\dfrac{4}{3}$ instead of making a substitution like we did before.

$$\dfrac{4}{3} \times \dfrac{3x}{4} = 6 \times \dfrac{4}{3}$$

$$\dfrac{12x}{12} = \dfrac{6}{1} \times \dfrac{4}{3}$$

$$x = \dfrac{24}{3}$$

x = 8

It may be easier now to show you how to model and solve exercise 6) $\dfrac{3x}{4} + 3 = 9$

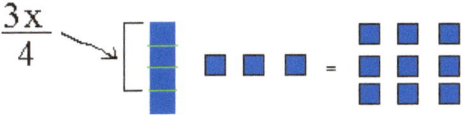

Or

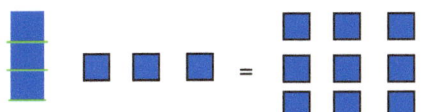

Add 3 red squares to both sides

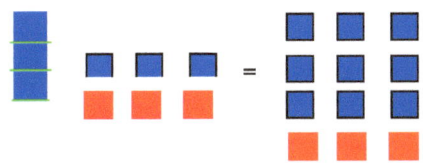

Perform cancelations of red squares with blue squares.

Now ask yourself, "How do I turn $\frac{3x}{4}$ into x?

It is kind of subtle! That is why I do not recommend modeling when the equation has fractions in it.

Notice that all you need to do is to attach the following piece or $\dfrac{1x}{4}$

Notice also that the piece above is equivalent to 2 blue squares. This is demonstrated below.

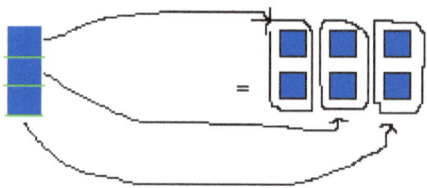

Therefore, after you add $\dfrac{1x}{4}$ to the left side, you will need to add 2 blue squares to the right side.

As you can see again

x = 8

## Summary and key exercises

a) An equation has a right side and a left side. The equal sign separates the left side from the right side.

b) Model the equation with rectangle(s) and square(s). 1 blue rectangle represents x. 2 blue rectangles represent 2x, and so forth…

c) Isolate 1 blue rectangle

d) Whatever you do to a side, do it to other side.

e) A red square next to a blue square cancel each other

1) -2 = x + 4

   Model the equation

   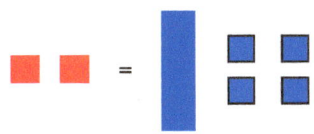

   Isolate the blue rectangle by getting rid of 4 blue squares. This can be done by Adding 4 red squares to both sides of the equation.

   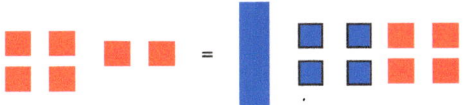

39

We get:

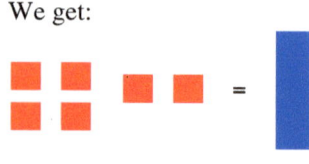

-6 = x or x = -6

You could also subtract 4 from both sides:

-2 = x + 4

-2 - 4 = x + 4 – 4
-6    = x

2) -2 = -x + 4

This exercise is exactly the same as exercise 1) with the exception that there is a negative next to x. After you do the exact same thing you did for exercise 1) you should end up with the following:

Turn the red rectangle into a blue rectangle. Then, following the principle that whatever you do to a side, you got to do it to the other side, you will also change all 6 red squares to 6 blue red squares.

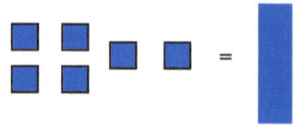

6 = x or x = 6

You could also subtract 4 from both sides and then multiply or divide both sides by a negative (-)

$-2 = -x + 4$

$-2 - 4 = -x + 4 - 4$

$-6 \phantom{xx} = -x$

Now, multiply both sides by a negative (-)

$- \times (-6) = - \times (-x)$

$\phantom{xx} 6 = x$

3)

$-3x + -4 = 8$

Model this equation

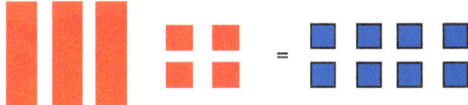

Isolate the 3 red rectangles by getting rid of 4 red squares. This can be done by Adding 4 blue squares to both sides of the equation.

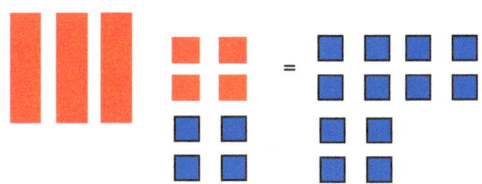

41

We get:

Isolate one red rectangle.

Turn the red rectangle into a blue rectangle. Then, following the principle that whatever you do to a side, you got to do it to the other side, you will also change all 4 blue squares to 4 red squares.

Thus, x = -4

You could also add 4 to both sides and then divide both sides by a negative (-3)

-3x + -4 = 8

-3x + -4 + 4 = 8 + 4

-3x    = 12

$$\frac{-3x}{-3} = \frac{12}{-3}$$

$$x = -4$$

4) $\frac{2x}{3} + 4 = 8$

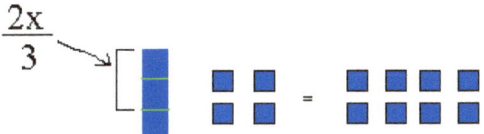

Or

Add 4 red squares to both sides.

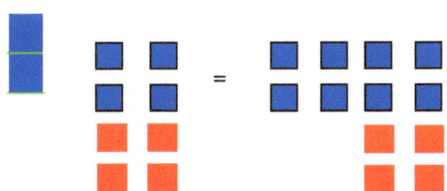

Perform cancelations of red squares with blue squares.

Now ask yourself, "How do I turn $\frac{2x}{3}$ into x?

Notice that all you need to do it to attach the following piece or $\frac{1x}{3}$

This piece is equivalent to 2 blue squares as shown below.

So when you add this piece to the left side to turn $\frac{2x}{3}$ into and x, you have to add 2 blue squares to the right side to keep the equation balanced.

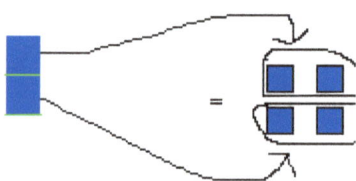

We get:

Or x = 6

You could also solve this equation by subtracting 4 from both sides.

Then multiplying both sides by $\dfrac{3}{2}$

$$\dfrac{2x}{3} + 4 = 8$$

$$\dfrac{2x}{3} + 4 - 4 = 8 - 4$$

$$\dfrac{2x}{3} = 4$$

$$\dfrac{3}{2} \times \dfrac{2x}{3} = \dfrac{3}{2} \times 4$$

$$\dfrac{6x}{6} = \dfrac{3}{2} \times \dfrac{4}{1}$$

$$\dfrac{6x}{6} = \dfrac{12}{2}$$

x = 6

_____Solving equations quiz_____

1) -4 + x = 5   _____

2) 5x = -15   _____

3) $\dfrac{1x}{3} = 6$   _____

4) 4x + 2 = 18   _____

5) $\dfrac{4x}{5} - 2 = 10$   _____

## Answers

_____ quiz_____

**1)** 9   **2)** -3   **3)** 18   **4)** 4   **5)** 15

# Unit3: Solving and graphing inequalities.

Main tool: A number line and background knowledge from unit 2.

Before showing you how to solve inequalities, let's get down some foundation. You will need to know how to graph the simplest situations or cases. Once you know that, everything else will be a breeze.

Case #1: Graph x > 3

The meaning of the graph above is to show on the number line all numbers bigger than 3. Refer to the graph below to see how we show this situation. Notice that 3 is not part of the graph. We show this with a circle that is **not** shaded.

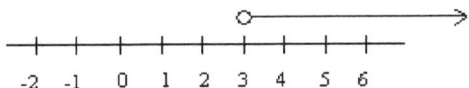

Case #2: Graph x ≥ 3

The meaning of the graph above is to show on the number line all numbers bigger or <u>equal</u> to 3. Refer to the graph below to see how we show this situation. Notice that 3 is part of the graph.

47

We show this with a circle that is shaded.

Case #3: Graph x ≤ -2

The meaning of the graph above is show on the number line all numbers smaller or equal to -2. Refer to the graph below to see how we show this situation.

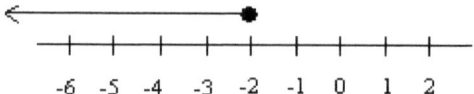

Case #4: Graph x ≠ -4

The meaning of the graph above is show on the number line all numbers except -4.

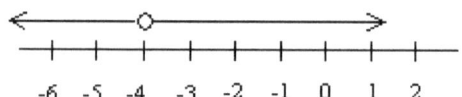

Notice again that -4 is not shaded.

## Another important concept:

We all know that $2 < 8$. Now, what Happens if we divide or multiply both sides by any negative number?
Choose -2 and divide both sides by -2

$$\frac{2}{-2} < \frac{8}{-2}$$

We get -1 < -4. Now ask yourself.

Is -1 really smaller than -4? Of course not!

-1 is bigger -4, so -1 > -4.

Now, multiply both sides by -2

2 × -2 < 8 × -2

-4 < -16

Again, since -4 is not smaller than -16, but bigger, change the direction of the inequality sign to >, so -4 > -16.

**Important concept to remember:**

Whenever you multiply or divide an inequality by a negative number, always change the direction of the inequality sign. This will be useful when we solve inequality and we have to divide or multiply by a negative number.

Now, let us solve some inequalities! Once you combined what we just learned here with the techniques to solve equations, you can solve any inequality. Why?
It is because we solve inequalities the exact same way we solve linear equations. We basically isolate x from both sides.

**Example #1:**

$2x + 8 > 14$

$2x + 8 - 8 > 14 - 8$

$2x + 0 > 6$

$2x > 6$

$\dfrac{2}{2}x > \dfrac{6}{2}$

$x > 3$

The graph is shown below.

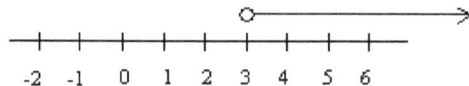

I can make one minor adjustment to example #1 by changing $>$ to $\geq$

Solve $2x + 8 \geq 14$

After you do the exact same thing, you will end up with $x \geq 3$. The graph is shown below.

**Example #2:**

$-3x + -4 \geq 2$

-3x + -4 + 4 ≥ 2 + 4

-3x + 0 ≥ 6

-3x ≥ 6

Now, we will divide both sides by -3. Just remember that when you divide both sides of an inequality by a negative number, you need to remember to change the direction of the inequality sign

$$\frac{-3}{-3}x \leq \frac{6}{-3}$$

$$x \leq -2.$$

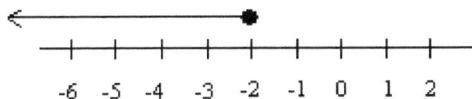

**Example #3:**

$$\frac{2x}{3} + 4 > 8$$

$$\frac{2x}{3} + 4 - 4 > 8 - 4$$

$$\frac{2x}{3} > 4$$

$$\frac{3}{2} \times \frac{2x}{3} > \frac{3}{2} \times 4$$

$$\frac{6x}{6} > \frac{3}{2} \times \frac{4}{1}$$

$$\frac{6x}{6} > \frac{12}{2}$$

x > 6

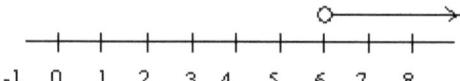

## Summary

Solving a linear inequality is very similar to solving a linear equation. There is one big difference. Whenever you divide an inequality by a negative number, switch the direction of the inequality sign or symbol.

# Unit 4: Exponent, roots, scientific notation, and The Pythagorean Theorem

## Exponent

*Exponent* is a shortcut for multiplication.

$7 \times 7 \times 7 \times 7 \times 7 = 7^5$

And $7^5$ means that 7 is multiplied by **itself** 5 times.

Common mistake to avoid: It does **not** mean $7 \times 5$.

In the same way, $7^3 = 7 \times 7 \times 7$

Now, what about $7^5 \times 7^3$ ?

$7^5$ means that 7 is multiplied by itself 5 times.

$7^3$ means that 7 is multiplied by itself 3 times.

Together, it means that 7 is multiplied by itself 8 times or $7^8$

$7^5 \times 7^3 = 7 \times 7 \times 7 \times 7 \times 7 \times 7 \times 7 \times 7$

$\phantom{7^5 \times 7^3} = 7^8$

Common mistake to avoid: It does **not** mean the exponent is going to be $5 \times 3$ or $7^{15}$ because there is a multiplication sign.

However, it does mean you can just add the exponent to get 8 as opposed to writing down the number 7 five times and then three times.

This will be useful if you are doing the following math problem:

$$4^{500} \times 4^{150}$$

You can just add 500 and 150 together to get 650, so

$$4^{500} \times 4^{150} = 4^{500+150} = 4^{650}$$

In general, $x^m \times x^n = x^{m+n}$

Now, try $\left(4^2\right)^3$

To really see what is going on here, we can make a useful substitution.

Let $y = 4^2$, so the expression becomes $(y)^3$ and $(y)^3 = y \times y \times y$

Replace y now with $4^2$ and the expressions becomes

$$4^2 \times 4^2 \times 4^2 = 4^6$$

54

Notice that you could get the 6 if you multiply 2 and 3. This will be very useful if

you are doing $\left(4^5\right)^{30}$

You would not want to write down $4^5$ thirty times.

Instead you can do 5 times 30 = 150 and the

answer is $4^{150}$

In general, $\left(x^n\right)^m = x^{n \times m}$

What if the problem is now $7^5 \div 7^3$ ?

$$7^5 \div 7^3 = \frac{7^5}{7^3} = \frac{7 \times 7 \times 7 \times 7 \times 7}{7 \times 7 \times 7}$$

$$= \frac{7 \times 7 \times 7}{7 \times 7 \times 7} \times \frac{7 \times 7}{1}$$

$$= 1 \times \frac{7 \times 7}{1}$$

$$= \frac{7 \times 7}{1}$$

$$= 7 \times 7$$

$$= 7^2$$

So, $7^5 \div 7^3 = 7^2$

You can get the number 2 if you just subtract 3 from 5. (5 - 3 = 2)

This will be useful if you are doing

$7^{884} \div 7^{84}$

Since 884 − 84 = 800, $7^{884} \div 7^{84} = 7^{800}$

If on the other hand the problem is

$7^{84} \div 7^{884}$ , we get:

$7^{84} \div 7^{884} = 7^{84-884} = 7^{-800}$

In general, $x^m \div x^n = \dfrac{x^m}{x^n} = x^{m-n}$

Try now, $4^5 \div 4^5$

**Observation #1:** Anything divided by the same thing is 1, so $4^5 \div 4^5 = 1$

**Observation #2:** $4^5 \div 4^5 = 4^{5-5} = 4^0$

We conclude from these two observations that

$4^5 \div 4^5 = 4^0 = 1$

In general, $x^0 = 1$

**Roots**

The best way to explain roots is to start with the *square root*.

The square root of 4 is 2 because 2×2 = 4

We write $\sqrt[2]{4} = \sqrt{4} = 2$

Notice that -2 also work since   -2x-2 = 4

We write $\sqrt{4} = -2$

So $\sqrt{4}$ has two answers. These are 2 and -2.

By the same token, $\sqrt{25}$ has two answers.

These are 5 and -5 because 5×5 = 25 and -5x-5 = 25

In general, $\sqrt[2]{x}$ written as $\sqrt{x}$ is a number a, such that   a×a = x

Just because we got two answers for $\sqrt{4}$ and $\sqrt{25}$ does not mean it is always the case for other roots.

What about the *cube root*?

The cube root of 27 is 3 because

3 ×3 × 3 = 27. We write $\sqrt[3]{27} = 3$

Here, there is only one answer. -3 ×-3 × -3 is not equal to 27.

57

What comes after the cube root? You can take the fourth root, fifth root, sixth root, seventh root, and so forth.

In general, $\sqrt[n]{x}$ is a number b such that

b × b × b × b × ... = x

The number of times you write down b depends on n.

If n = 5, then you will write b five times.

For example, what is $\sqrt[5]{1024}$ ?

Although this problem looks scary, it is very doable if you follow the definition. Here, you are looking for the fifth root.

According to the definition, you are looking for a number b such that

b × b × b × b × b = 1024

Don't try to guess. The most logical way to tackle this question is to start with 1. If 1 does not work, try 2, and so forth.

1 × 1 × 1 × 1 × 1 = 1
2 × 2 × 2 × 2 × 2 = 32
3 × 3 × 3 × 3 × 3 = 243
4 × 4 × 4 × 4 × 4 = 1024

Therefore, $\sqrt[5]{1024}$ = 4.

### Scientific notation

A number is in *scientific notation* if it has the following format:

$a \times 10^n$

a is at least 1 and less than 10.

n is an integer.

Let's put 9000000 in scientific notation.

$9000000 = 9 \times 1000000$

Since $10 \times 10 \times 10 \times 10 \times 10 \times 10 = 1000000$

$9000000 = 9 \times 10 \times 10 \times 10 \times 10 \times 10 \times 10$

$$= 9 \times 10^6$$

There is a better and more straightforward way to put a number in scientific notation.

Let us start again with 9000000

Any whole number always has a decimal point at the end. I show it in bold below.

So, 9000000 = 9000000**.**

Then, move this decimal point to the left until it is between 9 and 0.

The number of moves that you make is n

59

| | |
|---|---|
| 900000.0 | 1 move to the left |
| 90000.00 | 2 moves to the left |
| 9000.000 | 3 moves to the left |
| 900.0000 | 4 moves to the left |
| 90.00000 | 5 moves to the left |
| 9.000000 | 6 moves to the left |

Since I made 6 moves, n = 6.
Notice that I did not stop after 5 moves because I want the number to be less than 10. However, if I stop after 5 moves, the number will be 90 and 90 is not less than 10.

Again, you get $9000000 = 9 \times 10^6$

**Example #1:**

Put 570000000 into scientific notation.

570000000 = 570000000.

Then, move this decimal point to the left until it is between 5 and 7.

| | |
|---|---|
| 57000000.0 | 1 move to the left |
| 5700000.00 | 2 moves to the left |
| 570000.000 | 3 moves to the left |
| 57000.0000 | 4 moves to the left |
| 5700.00000 | 5 moves to the left |
| 570.000000 | 6 moves to the left |
| 57.0000000 | 7 moves to the left |
| 5.70000000 | 8 moves to the left |

Since you made 8 moves, n = 8

You get $5.8 \times 10^8$

What if you had 0.0000008? This time, the point is shown, so you will not put it at the end as you did before. Only when you don't see it, you can put it at the end.

While 80000000 = 8×10000000

$$0.0000008 = \frac{8}{10000000}$$

For 80000000, just count the 0 and that is how many zeros you put after the number 1 in the multiplication.

For 0.0000008, just count how many digits you have after the decimal point. That is how many zeros you put after the number 1 in the division

$$0.0000008 = \frac{8}{10000000}$$

$$\frac{8}{10000000} = \frac{8}{10^7}$$

We will now try to get rid of the denominator that is $10^7$.

The way to do this is to turn $10^7$ into a 1. This is demonstrated next.

$$\frac{8}{10000000} = \frac{8}{10^7} = \frac{8}{10^7} \times \frac{10^{-7}}{10^{-7}}$$

$$= \frac{8 \times 10^{-7}}{10^7 \times 10^{-7}}$$

$$= \frac{8 \times 10^{-7}}{10^7 \times 10^{-7}}$$

$$= \frac{8 \times 10^{-7}}{10^{7+-7}}$$

$$= \frac{8 \times 10^{-7}}{10^0}$$

$$= \frac{8 \times 10^{-7}}{1}$$

$$= 8 \times 10^{-7}$$

But again, there is a better and more straightforward way to put a number in scientific notation. Let's do it again!

0.0000008

Move the decimal point to the **right** this time after the number 8.

The number of moves that you make is n

| | |
|---|---|
| 00.000008 | 1 move to the right |
| 000.00008 | 2 moves to the right |
| 0000.0008 | 3 moves to the right |
| 00000.008 | 4 moves to the right |
| 000000.08 | 5 moves to the right |
| 0000000.8 | 6 moves to the right |
| 00000008. | 7 moves to the right |

I made 7 moves. The only difference here is that n is a negative number. So n = -7.

Notice that I did not stop after 6 moves because I want the number to be at least one However, if I stop after 6 moves, the number will be 0.8 and 0.8 is less than 1.

$$0.0000008 = 8 \times 10^{-7}$$

**Example #2**: Put 0.00000641 into scientific notation.

Move the decimal point to the **right** this time after the number 6.

The number of moves that you make is n

| | |
|---|---|
| 00.0000641 | 1 move to the right |
| 000.000641 | 2 moves to the right |
| 0000.00641 | 3 moves to the right |
| 00000.0641 | 4 moves to the right |
| 000000.641 | 5 moves to the right |
| 0000006.41 | 6 moves to the right |

I made 6 moves and n is negative. So n = -6.

$$0.00000641 = 6.41 \times 10^{-6}$$

### The Pythagorean Theorem

Consider the following triangle with the length of the hypotenuse equals to c and legs equals to a and b.

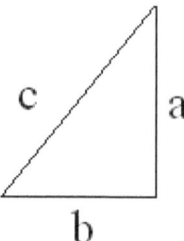

The Pythagorean formula states that

$$c^2 = a^2 + b^2$$

The Pythagorean formula helps you to get the third side when two sides are known.

**Exercise #1:** Find c when a = 3 and b = 4

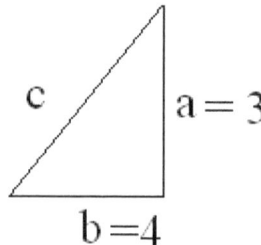

$c^2 = a^2 + b^2$

$c^2 = 3^2 + 4^2$

$c^2 = 9 + 16$

$c^2 = 25$

What number raised to the second power equals 25? Since $5^2 = 25$, c = 5.

**Notice that to get 5, you can take the square root of 25**

**Exercise #2:**

Find a when c = 10 and b = 8

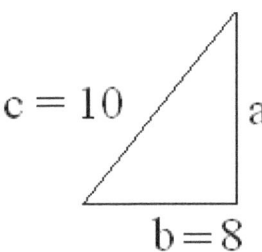

$c^2 = a^2 + b^2$

$10^2 = a^2 + 8^2$

$100 = a^2 + 64$

$100 - 64 = a^2 + 64 - 64$

$36 = a^2 + 0$

$36 = a^2$

What number to the second equals 36?

Since $6^2 = 36$, a = 6

## Summary and key exercises

$$x^m \times x^n = x^{m+n}$$

$$\left(x^n\right)^m = x^{n \times m}$$

$$x^m \div x^n = \frac{x^m}{x^n} = x^{m-n}$$

$\sqrt[n]{x}$ is a number b such that

b × b × b × b × ... = x

b is multiplied by itself <u>n</u> times

A number is in scientific notation if it has the following format:

$$a \times 10^n$$

a is at least 1 and less than 10.
n is an integer.

The Pythagorean formula states that

$$c^2 = a^2 + b^2$$

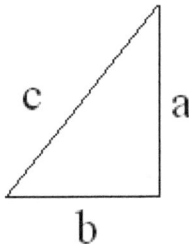

1)

Simplify $\dfrac{\left(4^3\right)^2 \times 4^8}{4^{19}}$

$$\dfrac{\left(4^3\right)^2 \times 4^8}{4^{19}} = \dfrac{4^{3\times 2} \times 4^8}{4^{19}}$$

$$= \dfrac{4^6 \times 4^8}{4^{19}}$$

$$= \dfrac{4^{6+8}}{4^{19}}$$

$$= \dfrac{4^{14}}{4^{19}}$$

$$= 4^{14-19}$$

$$= 4^{-5}$$

2) Write $4^{-5}$ with a positive exponent

$$4^{-5} = \frac{4^{-5}}{1} = \frac{4^{-5} \times 4^5}{1 \times 4^5}$$

$$= \frac{4^{-5+5}}{4^5}$$

$$= \frac{4^0}{4^5}$$

$$= \frac{1}{4^5}$$

3) Write $x^{-n}$ with a positive exponent

$$x^{-n} = \frac{x^{-n}}{1} = \frac{x^{-n} \times x^n}{1 \times x^n}$$

$$= \frac{x^{-n+n}}{x^n}$$

$$= \frac{x^0}{x^n}$$

$$= \frac{1}{x^n}$$

4)

Put the following numerical expression into scientific notation. $0.00004 \times 60000000$

$$0.00004 \times 60000000 = 4 \times 10^{-5} \times 6 \times 10^{7}$$
$$= 24 \times 10^{-5+7}$$
$$= 24 \times 10^{2}$$
$$= 2.4 \times 10^{1} \times 10^{2}$$
$$= 2.4 \times 10^{3}$$

5)

Using $c^2 = a^2 + b^2$ find b when a = 5 and c = 13.

$c^2 = a^2 + b^2$

$13^2 = 5^2 + b^2$

$169 = 25 + b^2$

$169 - 25 = 25 - 25 + b^2$

$144 = b^2 + 0$

$144 = b^2$

What number to the second equals 144?

Since $12^2 = 144$, b = 12

## Final test

1) 50 + -2 _____

2) -50 — - 9 _____

3) -50 — 9 _____

4) -50 ÷ 10 _____

5) -50 × -10 _____

6) x + -2 = 8 _____

7) 4x = 12 _____

8) $-\frac{2}{3}x = 6$ _____

9) 2x + 5 = 17 _____

10) -2x + 5 = 17 _____

11) $-\frac{1}{2}x + 3 = 7$ _____

12) $-2x + 5 \leq 17$ _____

13) $3x - 5 \leq -17$ _____

14) $-\frac{1}{2}x + 3 \geq 7$ _____

15) Put the following number in scientific notation.

   a. 0.0000000000000456
   b. 24000000000000000

16. Put the following number in standard form.

   a. $6.5 \times 10^8$
   b. $6.5 \times 10^{-8}$

17) $\dfrac{\left(x^4\right)^3 \times x^8}{x^{24}}$

18) Find the fifth root of 7776

19) Using the formula $c^2 = a^2 + b^2$

   Find c when a = 12 and b = 5

20) Using the formula $c^2 = a^2 + b^2$

Find a when b = 15 and c = 39

## Answers

_____Final Test_____

**1)** 48   **2)** -41   **3)** -59   **4)** -5   **5)** 500

**6)** 10   **7)** 3   **8)** -9   **9)** 6   **10)** -6

**11)** -8   **12)** $x \geq -6$   **13)** $x \leq -4$

**14)** $x \leq -8$   **15)** a. $4.56 \times 10^{-14}$

b. $2.4 \times 10^{16}$

**16)** a. 650000000   b. 0.000000065

**17)** $x^{-4}$   **18)** 6   **19)** 13   **20)** 36

www.ingramcontent.com/pod-product-compliance
Lightning Source LLC
Chambersburg PA
CBHW040834180526
45159CB00001B/191